TRAINS AROUND DONCASTER

JOHN JACKSON

AMBERLEY

First published 2024

Amberley Publishing
The Hill, Stroud
Gloucestershire, GL5 4EP

www.amberley-books.com

Copyright © John Jackson, 2024

The right of John Jackson to be identified as the Author of this work has been asserted in accordance with the Copyrights, Designs and Patents Act 1988.

ISBN 978 1 3981 1779 2 (print)
ISBN 978 1 3981 1780 8 (ebook)

All rights reserved. No part of this book may be reprinted or reproduced or utilised in any form or by any electronic, mechanical or other means, now known or hereafter invented, including photocopying and recording, or in any information storage or retrieval system, without the permission in writing from the Publishers.

British Library Cataloguing in Publication Data.
A catalogue record for this book is available from the British Library.

Typesetting by SJmagic DESIGN SERVICES, India.
Printed in the UK.

Contents

Introduction	4
Setting the Scene	8
Britain's Railways – From Nationalisation to Privatisation	11
Passenger Services – LNER	15
Passenger Services – LNER's Competitors	24
Passenger Services – Northern	28
Passenger Services – TransPennine Express (TPE)	37
Passenger Services – East Midlands Railway (EMR)	40
Freight Operators – DB Cargo	43
Freight Operators – Freightliner	54
Freight Operators – GB Railfreight (GBRf)	59
Freight Operators – Direct Rail Services (DRS)	67
Freight Operators – Colas Rail	73
Other Train Operators	78
Network Rail and Infrastructure	80
A Variety of Visitors to West Yard and Wabtec	84
Hitachi Movements on ECML	92
To the North East and South West	94
LNER and InterCity Livery	96

Introduction

My first visits to Doncaster station date back to the late 1950s and early 1960s when, whilst I was still a child, our family regularly changed trains there to visit relatives near Hull. It was the age of steam then, of course, with the 'Deltic' diesel locomotives just starting to make an appearance on East Coast Main Line (ECML) services. Although many of these earliest memories predate even my oldest trusty notebooks, I still recall one of my first sightings of the iconic class D9007 *Pinza* passing through the station non-stop on 'The Flying Scotsman', that morning's 10.00 from Edinburgh Waverley to London King's Cross.

I used to relish the challenge of seeing just how many numbers my father and I could jot down when passing Doncaster's steam shed, coded 36A. I can vividly recall the presence of A1 'Pacific', No. 60122 *Curlew*, on the steam shed one summer's day in the early 1960s. Sadly, I have no photographic mementos from those halcyon days.

The rail scene has, of course, changed dramatically in the ensuing half century and more, with electrification of the ECML around 1990, and rail privatisation a few years later, just two of the biggest and most wide-sweeping developments. The steam era has long gone, as have the 'Deltic' diesel locomotives. Nevertheless, the station area still draws enthusiasts to it like a magnet and it remains something of a railfan's hotspot. In this publication I set out to demonstrate why Doncaster retains its appeal in the twenty-first century.

There has always been a rivalry between the two major rail lines linking London with the Scottish cities of Glasgow and Edinburgh. The West Coast Main Line was constructed from London's Euston station and reached Scotland via Rugby, Crewe, Preston and Carlisle. Its rival was the East Coast Main Line from London's King's Cross station via Peterborough, York and Newcastle. Just over 155 miles north of London, the ECML passes through the major South Yorkshire railway junction at Doncaster.

A glance at the rail map on p. 7 will reveal the web of connecting lines in and around the Doncaster area. In particular, the busy route between London and West Yorkshire warranted electrification north-westwards to Wakefield and Leeds, around the same time as 'the wires' reached York and beyond. In addition, the station enjoys regular diesel-powered connections eastwards on both the north and south banks of the River Humber to Hull, Grimsby and Cleethorpes. In the westerly direction, there are regular diesel services to Sheffield and Manchester. Today's passenger picture is completed by an alternative, and much slower, service to Peterborough via Lincoln Central, operated by East Midlands Railway (EMR).

Both passenger operators' ownership and branding has changed regularly since the rail privatisation days of the mid-1990s, including franchises returning to government control. In addition to the EMR services mentioned above, six other operators' trains pass through the station. The flagship, long-distance services between London, Leeds and Edinburgh are chiefly in the hands of LNER. Regional services in the north of England are operated by both Northern and TransPennine Express. Hull Trains handle services between the Yorkshire city and London, with Grand Central also serving London from both Bradford and Sunderland. Finally, CrossCountry Trains currently provide a small number of services from the North East to Banbury and Bristol via Birmingham New Street. Most of the company's services on this axis currently go via the more populous route of Leeds and Wakefield, and not Doncaster, on journeys between York and Sheffield, despite the Doncaster route offering quicker timings. In the pages that follow we take a look at Doncaster's nine-platform station and each of these passenger operators' services in more detail.

The passenger framework, however, is not the primary motivation for enthusiasts visiting Doncaster. The area may have suffered more than most from the substantial downturn in railborne coal movements, but there remains considerable freight train activity in the immediate area, even though the station itself is bypassed by many workings which utilise the station's avoiding lines. The freight traffic operated by the country's three largest freight operators, DB Cargo, Freightliner and GB Railfreight, can be seen regularly in the proximity of the station, with less frequent appearances by Colas Rail and Direct Rail Services completing the picture.

In addition, to the south of the station lie both Decoy and Belmont yards, on both sides of the ECML. The area also houses Doncaster Railport and an LNER depot, Doncaster Carr, for the company's current passenger fleet of 'Azumas' to be serviced by their manufacturer, Hitachi. To the west, Electromotive Diesel has a servicing depot at Roberts Road, built adjacent to the Sheffield line. As a result, GB Railfreight Class 66 locos, and occasionally other classes and operators, stable in the area.

The city of Doncaster has long had a historic association with the nation's railways. As far back as the 1850s, the Great Northern Railway established its workshops in the town, becoming known locally as 'The Plant'. In the following 100 years, over 2,200 steam locomotives were built at the workshops, including probably the most well-known steam loco in the world, No. 60103 *Flying Scotsman*. As I write these notes, the A3 'Pacific' is approaching its 100th birthday, having been built in 1923 to a design by Sir Nigel Gresley. 'The Plant' was also responsible for building the A4 'Pacific', No. 60022 *Mallard*. This steam locomotive retains its place in history having reached a speed of 126 mph in 1938. This remains the world speed record for a steam loco to this day and *Mallard* features towards the end of this book.

By the early 1960s, the workshops had been responsible for overhauling around 10,000 steam locomotives. They were to enjoy a new lease of life as British Rail Engineering Ltd, both building and servicing diesel and electric locomotives. More recent examples of their construction include a total of eighty-five examples of the Class 56 locomotives, as well as the entire class of fifty Class 58 locos.

By 2000, the site had been acquired by Westinghouse Air Brake Company and rebranded as Wabtec. Now occupying a much smaller area, the company continues to overhaul and repair rolling stock. Its yard is partially visible from the station platforms, alongside the main West Yard, which remains home to a wide variety of stock to this day.

In this publication we take a detailed look at the twenty-first-century railway scene in and around Doncaster station, ranging from the transition into privatisation in the mid-1990s through to the present day. Starting with a look at the various passenger companies operating to and through the station, we then take a similar look at the freight operations. Finally, we highlight the range of visiting locos and rolling stock that are seen at both the station's West Yard and nearby Wabtec workshops.

Finally, I hope you enjoy the journey through the pages that follow as much as I have enjoyed compiling them.

John Jackson

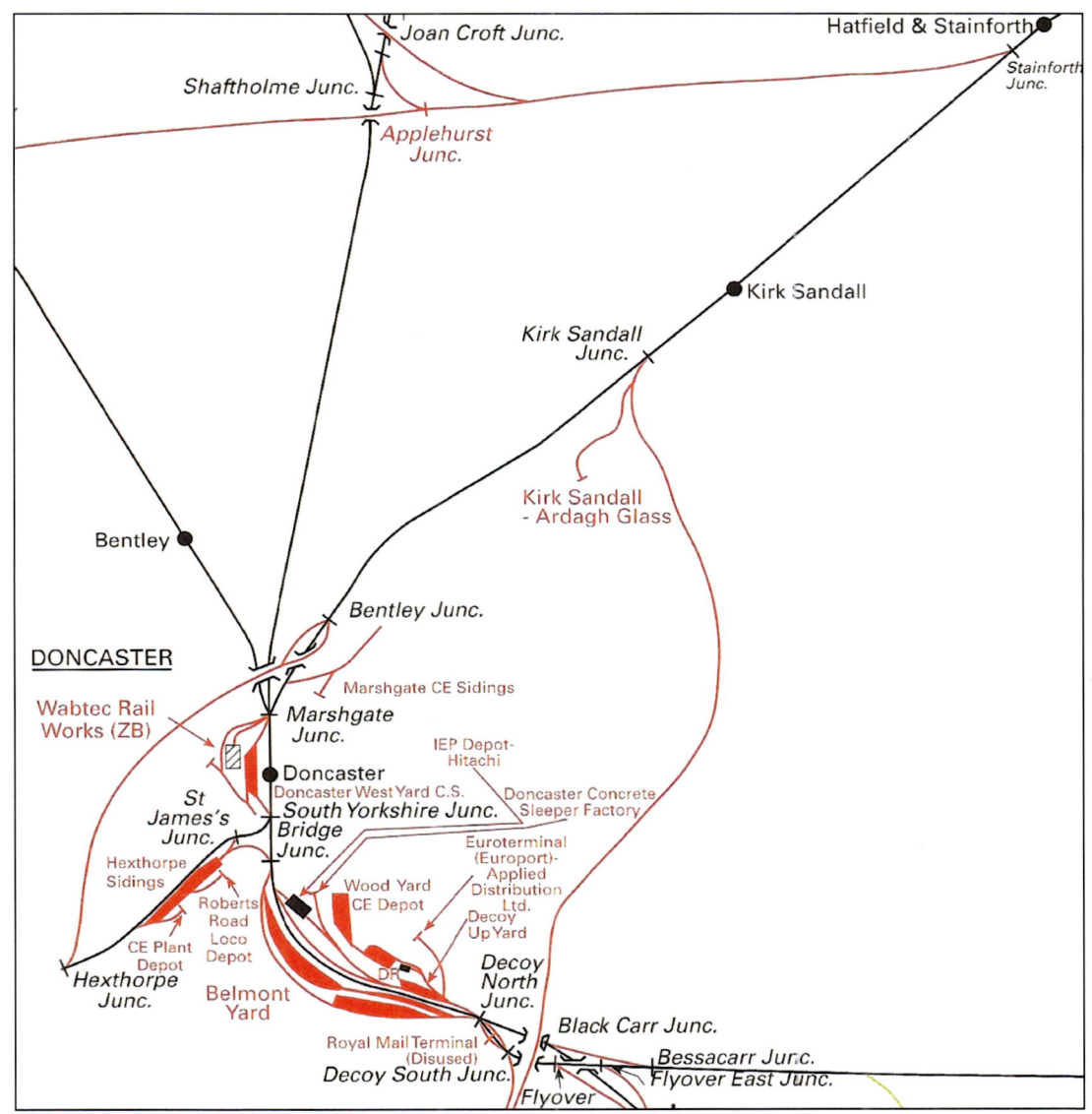

Rail map showing the lines in and around Doncaster.

Setting the Scene

This photo, taken in January 2023, shows the entrance to the city's station today. It was rebuilt in its present form at the turn of the Second World War, with minor modifications since. In 2006, the station was incorporated into the nearby interchange and Frenchgate shopping centre. These buildings can be seen to the right of the picture.

The Frenchgate building offers an aerial view of the station and surrounding area. This photograph shows the station on 7 June 2018, with Nos 67003 and 91120 both visible in the West Yard beyond. Until recently, the station consisted of two main islands with platforms 1 to 4 on the left and 5 to 8 on the right.

Beyond the station platforms and West Yard is the former Doncaster Works area, which forms 'Wabtec' today. The site had been acquired by Westinghouse Air Brake Company at the turn of the millennium and rebranded Wabtec. In this view on 6 July 2018 their resident shunters, Nos 08669 and 08724, can be seen in the right foreground.

In order to increase capacity, at the end of 2016 a ninth platform, platform 0, was added at Doncaster station. It is a north-facing terminating platform and used by trains that veer eastwards from the city, serving Hull in particular. The connecting footbridge linking it to the remainder of the station can be seen in this view taken a few weeks after opening. On 26 January 2017, No. 158787 waits to leave on a service to Hull and Beverley.

With Doncaster station platforms on the left, this is the view of West Yard looking south on 19 October 2015. Unit No. 322484 is occupying one of the sidings, awaiting transfer onwards to Wabtec's own yard.

Britain's Railways – From Nationalisation to Privatisation

This book features trains in and around Doncaster in the twenty-first century. The next few photos are a reminder of Doncaster in the mid-1990s when rail privatisation was 'in the pipeline'. The Class 91 electric locomotives were the last class to be acquired by British Rail for its flagship East Coast Main Line (ECML) route out of London's King's Cross station. On 29 July 1995, No. 91014 (later renumbered to No. 91114) calls at Doncaster on an InterCity service to London.

The InterCity-liveried High Speed Trains also worked long-distance services on the ECML. On 24 May 1994, power car No. 43013 is seen in platform 4 on a Doncaster call.

Class 90 electric locomotives were also in use on passenger services in the mid-1990s. On 3 August 1993, No. 90025, with Railfreight Distribution vinyls, is seen at the same platform awaiting departure to the north.

The Rail Express Systems (RES) colours were still in evidence across the network, particularly on the ageing fleet of Class 47 locomotives, many with names relevant to the RES branding. On 29 July 1995, No. 47765 *RESsaldar* is seen stabled in Doncaster's West Yard.

In that transition era in the mid-1990s the ever-growing fleet of Class 66 locomotives were yet to be seen. Doncaster station saw a substantial volume of coal traffic passing through, utilising the railways' extensive fleet of Merry-Go-Round (MGR) wagons. In August 1993, No. 56071 heads north on one such rake, bringing back memories of trying to 'capture' the numbers of forty or more wagons into a Dictaphone as the train passed!

After weekend stabling out of use, a convoy of Class 56 locomotives pass light engine through the station on a Monday morning in August 1992. The quartet of Nos 56030, 56081, 56069 and 56082 are about to resume coal duties in the Yorkshire area.

The substantial fleet of Class 37 locomotives was also in its prime. On 4 May 1994, 37698 heads south through Doncaster on a southbound tank working.

A couple of years earlier, in August 1992, sister loco No. 37032 passes through Doncaster on the centre road on a light engine working.

Passenger Services – LNER

Doncaster is an important stop for most long-distance services linking London King's Cross with West Yorkshire, north-east England and Edinburgh. In today's privatised era, these ECML services have been operated by a number of franchisees, including today's incumbent, LNER. On 28 November 2019, No. 91113 heads north on a non-stop service.

Most of these LNER services call at Doncaster, endorsing its status as a major interchange with other rail services. On 2 July 2013, No. 91101 calls on a southbound service before continuing its journey of a further 155 miles to London King's Cross. This loco was chosen to receive the impressive 'Flying Scotsman' branding, a name that will forever be synonymous with the ECML.

Unlike the nationalised era of the 'Deltic' class of locomotives, these services operate with a Driving Vehicle Trailer (DVT), thus avoiding a need for engine changes at terminating stations. Usually, the Class 91 is found at the north, or country, end of the train with the DVT at the south, or London, end. On 13 May 2013, DVT No. 82230 is seen on the south end of a Newcastle-bound working. It is sporting 'East Coast' livery.

Loco No. 91125 was chosen to carry a promotional livery for Sky 1 HD. Together with a matching rake of Mark IV coaching stock, it is seen making a call at Doncaster on 8 January 2014. The Class 91 is propelling its services towards London King's Cross.

Sister loco No. 91132 is seen calling at Doncaster on 22 January 2019. In addition to its LNER livery, the loco is also branded with the 'let's end mental health discrimination' pledge. The loco itself, originally numbered 91023, was seemingly jinxed after two serious accidents, resulting in its renumbering.

Class 90 locomotives were often hired in from other private operators to complement the Class 91 loco fleet. They were usually found on services from London King's Cross to Leeds or York. On 28 June 2018, No. 90039 approaches Doncaster station on 14.39 service to Leeds.

The iconic High Speed Trains (HSTs) also operated long-distance services alongside the Class 91s. They had long been the choice of diesel traction for the non-electrified lines, particularly in the north of Scotland. On 5 June 2013, power car No. 43238 leads an HST rake north through Doncaster. The train is the daily 'Highland Chieftain' working from London King's Cross to Inverness, which does not call at Doncaster.

Virgin are amongst the companies to have operated services on the ECML since rail privatisation in the mid-1990s. Their livery can be seen on power car No. 43208 *Lincolnshire Echo* as it hauls another Anglo-Scottish service southwards to King's Cross. Sister power car No. 43316 brings up the rear on 14 June 2017.

Whilst space precludes a full chronology of ECML operators since privatisation, here are two more examples of the liveries applied to the HST power cars. First, this 2012 view shows the East Coast livery and branding on power car No. 43312. On 13 October that year, it is seen on the rear of an ECML northbound stopping service.

Secondly, often referred to as 'red stripe' livery, this is the temporary branding applied by Virgin to its power car No. 43307. On 19 October 2015, it is at the rear of a southbound service standing at platform 3. This platform had by then been split into an 'A' and a 'B' end with this London-bound working using the 'A' end.

Virgin Trains East Coast (VTEC) often called upon East Midlands Trains to hire in additional HST power cars to cover their services. On 23 June 2017, EMT power cars, with No. 43059 leading and No. 43061 on the rear, make the Doncaster call on a King's Cross to Leeds service.

Two years later, on 14 March 2019, No. 43061 is seen again calling at the same platform 4 on the rear of another London to Leeds service. This time its EMT partner is No. 43075 leading the service.

Maintaining the fleet of ECML locomotives often sees additional light-engine moves for operational purposes. On 13 February 2018, for example, hired-in No. 90034 arrives at Doncaster. It has dragged Nos 91005 and 91020 from the depot at Bounds Green in north London. It will shortly recess the pair into the adjacent West Yard.

For a number of years, DB Cargo Class 67 locomotives have provided a 'Thunderbird' service should rescue of a failed train on the ECML become necessary. On 17 January 2017, DB Cargo's No. 67024 has been called upon to move No. 91120 from Bounds Green to Doncaster.

As any rail enthusiast visiting Doncaster will tell you, two trains arriving at once is an occupational hazard. In this view on 30 June 2017, GB Railfreight's No. 66772 is about to head north through platform 4. Approaching at speed on the Down main line is VTEC's No. 43306 on a northbound Anglo-Scottish service.

The reign of HSTs on ECML services, together with the majority of Class 91 locos, was coming to an end. They were shortly to be displaced in a new era as the Azuma was about to make its debut. On 8 February 2017, No. 800101 is seen in platform 1 on a test working. The imminent Virgin Azuma services claimed 'more seats, more service, more style'.

These Azumas, the Japanese word for 'east', were built with Japanese bodyshells and assembled at Hitachi's UK plant at Newton Aycliffe, County Durham. On 30 August 2019, No. 800105 rushes past Doncaster on the Up main line as it heads towards London King's Cross.

By the time this photo was taken in 2021 the Azumas were operating the majority of LNER services in and out of London King's Cross. On 4 August that year, No. 800103 is seen approaching platform 4 on a service from London. Two of the now redundant Class 91 locos can be seen in West Yard to the right. At the time of writing, LNER is in the hands of the DOHL, a public corporation of the Department for Transport. They stepped in to maintain the continuity of passenger services following a franchise termination.

Passenger Services – LNER's Competitors

The various ECML operators have encountered competition for customers on services northwards from Doncaster and York. CrossCountry Trains operate a number of services that offer the traveller an alternative service to Newcastle and Scotland, using their fleet of Voyagers. On 21 May 2014, No. 220014 has just joined the ECML at Doncaster on a northbound service to Newcastle.

On 20 February 2020, Voyager No. 221144 is seen calling at platform 3B on a southbound service from Newcastle. It will now leave the ECML and head west towards Sheffield, then on to Birmingham and beyond.

Further competition arrived in the post-privatisation era when, in 2000, open railway access was granted to Hull Trains to operate services between the East Yorkshire city and London King's Cross, via Doncaster. On 6 September 2018, Class 180 'Adelante' unit No. 180111 calls at Doncaster on a London to Hull service.

These Class 180 units replaced the Class 222 'Pioneer' sets on these services from 2008 onwards. On 13 May 2013, No. 180110 is seen approaching Doncaster from the south. Hull Trains is now wholly owned by FirstGroup and the access agreement is currently scheduled to run until 2032.

In 2015, Hull Trains decided to replace these Class 180 units with a new fleet of five Class 802 units, built by Hitachi. These were to become known as 'Paragon' units. Their introduction commenced in late 2019 and the service suspension during the Covid-19 pandemic resulted in the end of use of their Class 180 predecessors. On 4 August 2021, No. 802305 calls on a Hull-bound service.

In 2007, yet more competition arrived in the form of Grand Central Trains. After numerous unsuccessful bids, another open access agreement was granted to operate services between Sunderland and King's Cross. This was followed up by a similar agreement between Bradford and London around three years later. On 23 June 2017, two of their services pass at Doncaster, with HST power cars Nos 43423 and 43480 in view.

The company's use of HSTs has since ceased and all services are currently in the hands of Class 180 'Adelante' units. On 20 June 2019, No. 180101 powers through Doncaster on a London King's Cross to Sunderland service. Grand Central's open access agreement is in place until the early 2030s.

Yet more competition for Anglo-Scottish business via the ECML arrived in 2021 through a brand called 'Lumo'. Its name was derived from a combination of Luminosity and Motion. The company was to commission a new build of five Class 803 units operating between Edinburgh and London. With no permitted calls between Newcastle and Stevenage, these units pass through on Doncaster's centre roads. On 9 June 2022, No. 803005 heads north on one such service bound for Edinburgh.

Passenger Services – Northern

Northern Rail is another company currently operated by DOHL. Since 2020, this Department for Transport public corporation has been responsible for services across a vast swathe of the north of England, including around Doncaster. On 13 May 2013, one of Northern's much maligned 'Pacer' units, No. 142020, approaches Doncaster on a Sheffield to Adwick service.

Heading in the opposite direction on 22 January 2019, No. 142024 arrives in platform 3 on a service from Scunthorpe. It will cross the ECML here and head west on a stopping service to Sheffield, via Rotherham Central.

These 'Pacer' units have attracted a number of derogatory comments in the wider press. Originally introduced back in the mid-1980s, these Class 142 units clocked up more than thirty years' service before final withdrawal in 2020. On 6 December 2017, No. 142027 waits on the curve for the signal to approach Doncaster station on a Sheffield to Scunthorpe working.

Any 'Pacer' journey, no matter how short, is not a pleasant, comfortable experience but Northern's resources were stretched to the limits on 13 February 2018. Imagine the author's surprise whilst waiting to board Doncaster's 14.19 service to Bridlington to find that the unit allocated arriving from Sheffield was No. 142063. If it was to work throughout, that is a distance of approximately 113 uncomfortable miles. Thankfully, we have finally seen the end of Northern's fleet of Class 142s.

Northern's Class 144 'Pacer' units comprised a fleet of twenty further units built a couple of years after the Class 142s. They, too, were to be found on local services around Doncaster for more than thirty years. On 5 June 2013, No. 144001 stands in platform 2, a south-facing bay platform.

On 8 February 2015, sister unit No. 144007 is seen crossing the ECML through lines to enter platform 3B. The unit is forming a Sheffield to Scunthorpe stopping service. An hourly service, calling at all stations, has operated for most of this century.

Another Class 144, No. 144008, is seen emerging from the trackside vegetation as it approaches Doncaster from the Sheffield line. The unit is forming another all stations to Scunthorpe service on 19 October 2016.

One of these units, No. 144012, was chosen as an experimental variant in order to satisfy the requirements of 'Persons with reduced mobility' directives. It is seen here at Doncaster's platform 3B on 16 March 2018. This so-called '144 Evolution' project was not progressed and the fleet had been withdrawn by the end of 2020.

The stations between Sheffield and Doncaster are normally served by a half-hourly stopping service calling at all stations via Rotherham Central. Beyond Doncaster, one of these services is extended to and from Scunthorpe each hour, with the other making the short run to Adwick and back. On 20 June 2018, 'Sprinter' unit No.150110 has just returned to Doncaster's platform 8 on an Adwick to Sheffield service.

Another 'Sprinter' unit, No. 150136, is seen leaving Doncaster on 14 March 2019. The unit proudly displays the much-heralded 'Northern Powerhouse' and 'The Great North Rail Project' branding, although few tangible improvements have been delivered to date.

An alternative power formation was seen on this Sheffield to Scunthorpe service on 23 August 2018. A pair of single-car Class 153 units, Nos 153317 and 153370, were diagrammed. The latter unit, nearest the camera, still retained its Great Western Railway livery at the time.

Passenger Services – Northern

Northern operates an hourly service between Doncaster and Hull. These trains usually occupy the recently constructed platform 0. On 20 September 2017, No. 158790 is seen at this platform waiting to depart for Hull.

In addition to Northern's hourly Doncaster to Hull service, they also operate an hourly service between Sheffield and Scarborough, via Hull. This provides a half-hourly service for stations along the north bank of the Humber. On 8 March 2018, No. 158871 is working a Scarborough to Sheffield service. The unit still displays the colours of its former operator, ScotRail.

Several of ScotRail's Class 170 units have also been transferred to Northern. On 6 February 2019, No. 170455 stands in Doncaster's West Yard on one of these transfers.

Sister unit No. 170458 was already carrying Northern vinyls on 30 May 2019. It is seen calling at Doncaster on a Scarborough to Sheffield service.

Whilst LNER operate Doncaster to Leeds services calling only at Wakefield Westgate, Northern operate an hourly service along the route, calling at all stations. Until recently these services were operated from a small pool of electric multiple units (EMUs), including No. 321903. On 7 April 2012, it is seen arriving in platform 7, one of Doncaster's northern bay platforms, on a terminating service from Leeds.

Four-car EMUs are the usual motive power for these services. On 20 June 2019, a diesel multiple unit substitution was provided. Three-car unit No. 158755 is seen waiting in platform 6 to form the 12.26 departure to Leeds.

Since 2019, Northern have introduced a new class of EMU on these services. A small pool of twelve four-car Class 331 'Civity' units, built by CAF at Zaragoza in Spain, now operate these workings. On 9 June 2022, No. 331102 awaits departure from platform 6 on a Leeds service.

Northern has also taken delivery of their Class 195 'Civity' units, the diesel-powered equivalent of the electric Class 331s. During the class's testing phase, No. 195130 is seen stabled in Doncaster West Yard on 28 November 2019.

Passenger Services – TransPennine Express (TPE)

An east to west service has operated for many years linking Cleethorpes and South Humberside with Doncaster, Sheffield and Manchester. Some of these services were still in the hands of TPE's Class 170 units when this photo was taken in 2016. On 5 July that year, Nos 170307 and 170308 are paired on a service bound for Manchester Airport.

They were working alongside TPE's fleet of fifty-one Class 185 units. These three-car units form the majority of these South Pennine services. On 6 June 2014, for example, No. 185134 is leaving Doncaster on a westbound service.

A replacement TPE livery had been applied to the entire Class 185 fleet by the time this photo was taken on 23 August 2018. This 12.42 service from Doncaster to Manchester Airport had been delayed on its inbound journey from Cleethorpes. It was further delayed outside the station whilst waiting for a gap in other traffic in order to cross the busy ECML running lines. Eventual arrival for No. 185145 was at 13.05 and unusually required the use of platform 8 for its call at Doncaster.

TransPennine Express placed an order with Hitachi for a fleet of nineteen five-car bimode units. Although primarily intended to work services between Liverpool and Newcastle via Leeds, these units were seen through Doncaster whilst out on ECML test from Hitachi's plant at Newton Aycliffe. Pioneer unit No. 802201 is seen at Doncaster on 14 March 2019.

Class 68 locomotives and sets of Mark V coaching stock are intended to be introduced on Cleethorpes to Manchester services at some point in the future. In this connection, Class 68 loco No. 68031 *Felix* heads south through platform 3 in order to stable in Decoy Yard.

Passenger Services – East Midlands Railway (EMR)

Our review of Doncaster's scheduled passenger services is completed by EMR's local stopping service to Lincoln Central and Peterborough. Various operators have been responsible for these services since privatisation, including Central Trains. On 28 September 2004, their single car unit, No. 153385, waits at Doncaster's platform 5 on a service to Peterborough via Lincoln.

By 2017, these services were in the hands of East Midlands Trains. On 6 December that year, another single-car Class 153 unit, No. 153302, waits in the bay to form 13.01 service to Lincoln Central.

Despite Northern having a large fleet of Class 156 units, these sets made few appearances in the Doncaster area. By the time this Northern-liveried unit, No. 156470, appeared on 21 February 2012, it had already been transferred to East Midlands Trains the previous year. It is seen arriving on a terminating service from Lincoln.

By 2022, the Class 153 units had been withdrawn and services passed into the hands of EMR. Their reliveried two-car unit, No. 156907, arrives at Doncaster on 26 July that year.

Several different types of EMR's diesel units now appear on these services. On 16 March 2022, for example, Class 170 unit No. 170515 has just terminated at Doncaster.

The variety is completed by the company's use of Class 158 units. On 11 January 2023, it was the turn of No. 158864 to appear on this terminating service from Peterborough.

Freight Operators – DB Cargo

All the major freight operators' locomotives can be seen around the Doncaster area, including DB Cargo. As English, Welsh & Scottish Railways (EWS), it was one of the companies to benefit from the division of assets at the time of rail privatisation thirty years ago. Much has changed since then, including the introduction of a substantial fleet of General Motors-built Class 66 locomotives. On 8 February 2017, No. 66067 is seen stabled in Doncaster's Decoy Yard, to the south of the station. It was employed on yard shunting duties that day.

A year earlier, on 6 January 2016, it was the turn of veteran Class 60 locomotive, No. 60049, to be engaged on these shunting duties in Up Decoy Yard. This yard is to the east of the adjacent ECML, with its sister yard, Down Decoy, along with Belmont Yard, to the west side of the ECML. Both are visible from trains using the ECML and Lincoln lines.

These yard shunting duties were in the hands of Class 08 diesel shunters until their withdrawal by DB Cargo. On 23 February 2015, No. 08782 *Castleton Works* is employed on these duties. The move from wagonload freight trains to block works eliminated much of the need for splitting trains en route. Nowadays, the Decoy Yard is chiefly used for the formation of infrastructure workings in connection with National Rail possessions.

These shunting locomotives were stabled and serviced at the nearby locomotive depot. On 23 September 2012, DB Cargo shunter No. 09006 is seen stabled there. The depot was to close on 30 April 2014, and the site redeveloped for the maintenance of Hitachi's Azuma units for LNER.

On the opposite side of the ECML, No. 66077 stands at the north end of Belmont Yard on 1 March 2017. It is at the head of an infrastructure working ready to move northwards at the end of the day.

On 20 February 2020, No. 66113 heads north on the centre roads through Doncaster station. It has just left Up Decoy Yard on an infrastructure working to York Engineers Yard.

On 1 July 2020, No. 66017 is seen heading in the opposite direction through the station. It is hauling another lengthy infrastructure working on its return to Decoy Yard.

Also heading south, this time on 20 June 2019, No. 66134 is returning to Decoy Yard. It had been engaged on infrastructure duties earlier in that day and was now returning from York to Doncaster.

The freight movements in and around the Doncaster area were drastically reduced following the downturn in coal traffic, particularly given its position in the Yorkshire, Nottinghamshire and Derbyshire coalfield. DB Cargo HTA coal wagons were once a familiar sight through Doncaster, but not so today. On 5 June 2013, No. 66152 *Derek Holmes – Railway Operator* passes through the station on a local working. A landslip at Stainforth had resulted in a number of local diversions that day.

Sister loco No. 66012 was also on diversion that day. It had just run round its short rake of tanks and is heading back north through the station. The empty oil tanks were being worked from Neville Hill depot, Leeds, back to Lindsey Oil Refinery, on South Humberside.

On 8 January 2019, No. 66066 heads south on the two-way goods line alongside Doncaster station. It is returning a rake of empty wagons from Heck, North Yorkshire, to the quarry at Dowlow, near Buxton.

One of the few freight flows that has continued to utilise the Channel Tunnel since its opening in 1994 is the DB Cargo steel flow that links Scunthorpe, on South Humberside, with Ebange, in northern France. On 20 September 2017, No. 66115 heads north with a rake of empty steel wagons returning to Scunthorpe.

Another long-standing freight flow passing through Doncaster sees sand trains from Middleton Towers, near King's Lynn, to both Monk Bretton and Barnby Dun. In South Yorkshire, back in 2013, this traffic was handled by DB Cargo. On 13 May that year, No. 66172 is seen on a northbound working.

The loco depot at Doncaster may have closed in 2014, but DB Cargo locomotive stabling continues in Doncaster's Belmont Yard between their duties. This results in regular light-engine movements to and from Belmont. On 9 June 2022, for example, Nos 66099 and 66019 have just left Belmont heading for Scunthorpe.

On 4 August 2021, it is the turn of DB Cargo's Maritime liveried No. 66148 to head north from Belmont, this time to the Euro Terminal near Wakefield. It will work an intermodal from there to DP World's London Gateway freight terminal.

A more unusual DB Cargo working occurred on 10 February 2023. Their Class 66, No. 66221, was called upon to return three coaches from their depot at Knottingley to Nemesis Rail's complex at Burton upon Trent. It is seen on the centre road waiting for the signal to head westwards.

For many years, DB Cargo has provided Class 67 locomotives as 'Thunderbirds' in the event of a failure or emergency on ECML passenger services. These duties include stabling in Doncaster's West Yard. In this view, on 5 July 2016, Nos 67021 and 67003 can both be seen beneath the works' footbridge. LNER's No. 91127 can also be seen at the buffer stops.

On 6 September 2018, it is No. 67002 allocated to these 'Thunderbird' duties. The loco is seen in West Yard, sporting the livery of Arriva Trains Wales.

EWS-liveried No. 67016 is on standby for potential 'Thunderbird' action when seen in West Yard on 13 January 2022. Although they are infrequently called upon for their rescue duties, the author recalls this particular loco rescuing his Class 91-hauled ECML service in November 2015.

DB Cargo occasionally provides one of its 'royal' celebrities for these duties. On 3 October 2013, for example, No. 67005 *Queen's Messenger* stands in West Yard. Sister loco No. 67022 is to the left.

Alternative revenue-earning duties befell Chiltern-liveried No. 67012 on 8 January 2014. It was used to return a Chiltern Railways Driving Vehicle Trailer and one coach to Barton Hill depot in Bristol.

Freight Operators – Freightliner

The second operator of freight services in the Doncaster area, Freightliner, has been operating container traffic for thirty years since rail privatisation. Prior to that, the company had been set up, and was owned by, the British government in the late 1960s. On 3 August 2018, No. 66528 passes Doncaster West Yard with a northbound freightliner from Felixstowe to Leeds. By this date, the company was wholly owned by an American Railroad Holding Company, Genessee & Wyoming.

The company's small fleet of Class 70 locomotives are also occasionally seen on these freightliner services. On 29 July 2017, No. 70019 is the chosen motive power for that day's Felixstowe to Leeds service.

Heading in the opposite direction on 16 November 2016 is No. 66532. It is seen waiting alongside Decoy Yard at the head of this freightliner bound for Felixstowe, viewed from passing on a Doncaster to Lincoln passenger unit. The Freightliner will follow this unit towards Gainsborough Lea Road and reach the Suffolk port via Lincoln, Spalding and Peterborough.

Freightliner's 'Heavy Haul' division also handled a number of coal workings in and around Doncaster. On 2 July 2013, No. 66526 is seen heading north on Doncaster's through lines with a rake of coal empties. Sadly, the downturn in coal traffic has rendered scenes like this a thing of the past.

Thankfully, the company continues to handle a variety of other traffic in the Doncaster area. On 23 March 2023, for example, No. 66610 heads north on a working from Earles Sidings, in Derbyshire's Hope Valley, to the nearby power station at Drax.

Another power-station working on 10 November 2014 sees Freightliner's No. 70010 haul an empty rake of aggregate wagons through the station as it heads for Tunstead Quarry in Derbyshire. The working had originated from the power station at Eggborough, near Knottingley. This power station opened in the early 1960s but was closed completely in autumn 2018.

Freight Operators – Freightliner 57

On 21 August 2013, No. 66605 hauls another rake of empty hoppers south through Doncaster station. The wagons will be stabled in the yards to the south until their next duty.

Freightliner has a major servicing and maintenance depot at Midland Road/Balm Road in Leeds. This leads to frequent loco movements to and from Leeds, with many of these passing through Doncaster. On 29 September 2012, a quintet of the company's Class 66 locos heads south through the station platform to Decoy Yard, with No. 66515 leading.

Heading in the opposite direction on 6 September 2018 is sister loco No. 66623, named *Bill Bolsover* at the time. The loco is on its way to the company's Leeds depot from the yard at Whitemoor, in March, Cambridgeshire.

On 29 June 2017, No. 66951 has been despatched from Midland Road to Decoy Yard, Doncaster. It will pick up its wagons there before making the long journey to Fairwater Yard, near Taunton, Somerset.

Freight Operators – GB Railfreight (GBRf)

Formed after the mid-1990s rail privatisation, GBRf was regarded as one of 'the new kids on the block' when it commenced intermodal services in the UK. One of the first workings linked Potters' Logistics site at Selby, North Yorkshire, with the port of Felixstowe. On 17 September 2013, No. 66733 leads the southbound service to the Suffolk port.

Another southbound intermodal is seen passing through the station on 20 June 2019. This working from Tees Yard to Doncaster is hauled by No. 66789 *British Rail 1948–1997*.

The company has enjoyed steady growth in the last decade or so, with its expanding fleet of Class 66s now totalling around 100 locomotives, with Doncaster one of its major stabling and maintenance hubs. On 9 January 2015, No. 66766 stands on ElectroMotive's Roberts Road servicing depot, adjacent to the Doncaster to Sheffield line. Diesel shunter D4157 (No. 08927) stands alongside.

From the turn of the century, GBRf has also operated a variety of coal trains in the Yorkshire area. On 23 September 2012, No. 66703 heads a southbound rake of coal hoppers for stabling in Decoy Yard. Marking the twenty-first anniversary of the area's Power Signal Box, this locomotive is appropriately named *Doncaster PSB 1981–2002*.

As mentioned earlier, the power station at Eggborough, near Knottingley, closed completely in 2018, bringing an end to GBRf coal deliveries. On 6 November 2014, No. 66760 *David Gordon Harris* has brought a rake of empty coal wagons from the power station for stabling in Decoy Yard.

In common with other freight operators, GBRf has seen a drop in revenue as a result of the downturn in railborne coal traffic. One of the few surviving workings for the company is the flow between Immingham Docks and Ratcliffe Power Station. On 12 April 2022, No. 66708 *Glory to Ukraine* passes on the two-way goods line taking a rake of empties back to the docks at Immingham.

Some of the redundant coal hoppers have found use on other GBRf workings, such as aggregate traffic. On 20 June 2020, for example, No. 66709 is seen on a rake of these wagons heading south through Doncaster on a Rylstone Quarry, North Yorkshire, to Hull Dairycoates working.

Another working involving Rylstone is seen heading in the opposite direction on 28 June 2018, behind No. 66760 *David Gordon Harris*. The loco is taking a rake of empty wagons from GBRf's yard at Wellingborough, Northamptonshire, for loading at Rylstone.

Freight Operators – GB Railfreight (GBRf)

The yards in the Doncaster area are used to stable both locos and wagons, resulting in occasional shunting moves. On 11 April 2019, No. 66744 moves a single hopper wagon in the Roberts Road depot area.

To the south of the city and adjacent to the ECML, both the Decoy Yard and the former Royal Mail terminal sidings are used by GBRf. On 23 November 2018, No. 66749 stands in the former Royal Mail terminal, surrounded by a variety of the company's wagons.

In a competitive marketplace, GBRf continues to gain market share of the UK freight business. In some cases, that means winning business from freight competitors. The sand traffic from Middleton Towers to Barnby Dun had been handled by DB Cargo until GBRf took over at the beginning of January 2014. On 28 June 2018, No. 66701 is heading north to Barnby Dun, just a few miles to the east on the line towards Stainforth.

On 12 February 2023, two of GBRf's Class 66 locos were working to and from an engineers' possession at Whitehall Junction, close to Leeds station. At the head of the train, No. 66775 *HMS Argyll* is seen returning the wagons to Doncaster Decoy Yard.

Freight Operators – GB Railfreight (GBRf,

The two locos are working in 'top and tail' mode, with No. 66746 bringing up the rear of the train. For much of the year, this loco is more likely to be found on Royal Scotsman passenger duties.

To the west of Doncaster, locos leaving the yards join the Sheffield line at Hexthorpe. On 8 February 2017, No. 66778 waits at the junction to head west.

Another light engine move on 28 March 2018 involved No. 66783. The loco had recently been outshopped in a dedicated 'Biffa' livery. At the time of the photo, the nameplates were covered over. The loco was heading to York for a naming ceremony. It would later return displaying *The Flying Dustman* nameplates.

A more unusual GBRf visitor to Doncaster on 23 November 2022 was electro-diesel No. 73963 *Janice*. It is seen stabled in West Yard, believed to be about to receive attention within Wabtec.

Freight Operators – Direct Rail Services (DRS)

DRS locos are regular visitors to the Doncaster area on both engineers' infrastructure services and Network Rail test trains. On 20 June 2018, No. 66422 heads north through platform 3 on a working from Doncaster to Millerhill Yard, on the outskirts of Edinburgh.

A few weeks later, on 6 September, the same working was handled by Class 88 loco No. 88008 *Ariadne*. These electro-diesel locomotives are able to operate using the overhead line equipment for the majority of this journey.

Their diesel-only counterparts, the Class 68 locomotives, also appear on these infrastructure workings. On 6 September 2018, No. 68021 heads north on an infrastructure working from Doncaster to York.

On 24 May 2022, No. 66422 is seen working in the opposite direction. It is heading southbound on a working from Redcar to Doncaster Decoy Yard.

The southbound working from York on 29 June 2017 was a lengthy rake of wagons hauled by No. 66434. Several point carrier wagons were included at the front of the train.

The company's fleet of so-called heritage locos are slowly being phased out. On 23 August 2018, the nostalgia fraternity of rail enthusiasts was pleased with the choice of veteran machine, No. 57007, for this infrastructure working to York. This loco was built back in 1965.

These two locos from DRS' small fleet of Class 20s are even older having been built in 1961 and 1962 respectively. First, No. 20303 is seen stabled in the sidings to the south of the station on 24 March 2014.

Secondly, on 21 July the same year, No. 20309 accompanies No. 20308 on a light-engine move. The pair has arrived on a Crewe to Doncaster transfer.

The DRS Class 37s have been regular visitors to Doncaster for many years, mainly on Network Rail duties. On 22 March 2006, No. 37510 stands in West Yard between duties.

The Class 37s now see infrequent use on Network Rail test trains. In January 2016, No. 37667 stands in West Yard at the head of a short rake of the familiar yellow Network Rail rolling stock.

A very different working was handled by DRS in January 2018. A pair of their class 66 locomotives, Nos 66421 and 66429, were provided in connection with snowplough test runs. With a snowplough at either end, the pair are seen stabled in West Yard.

No less than three class members were present in West Yard on 14 September 2021. Locos Nos 66426, 66430 and 66432 are seen alongside DB Cargo 'Thunderbird' loco No. 67028.

Freight Operators – Colas Rail

The company now provides the motive power for the majority of loco-hauled test trains, on behalf of Network Rail. On 22 February 2020, No. 37254 stands in West Yard on a rake of test train stock.

Colas Rail made a popular choice for Class 37 rail enthusiasts when they decided to hire No. 37025 *Inverness TMD* for test-train duties. The large-logo, blue-liveried loco is owned by the Scottish 37 Group and is normally based at Bo'Ness. On 16 February 2018, the loco stands on a test train in Doncaster's West Yard.

Colas Rail has operated a diverse fleet of loco classes in recent years. This includes a pair of Class 67s, No. 67027 *Charlotte* and No. 67023 *Stella*. They are seen in platform 5 on 5 January 2018. This south-facing bay platform is more usually used for passenger services for Lincoln and Peterborough and, occasionally, Sheffield. The pair are reversing on an out and back light-engine working from Derby.

Another example of this diversity came on 8 September 2016 with No. 47739 *Robin of Templecombe* arriving light engine. The loco is awaiting entry to the Wabtec complex.

Freight Operators – Colas Rail

Back in 2013, Colas Rail had a small share of the UK's railborne coal traffic, with services from Wolsingham, on the Weardale Railway. These trains served Scunthorpe steel works and, a little later, the power station at Ratcliffe on Soar. On 21 June that year, No. 66848, one of Colas' small fleet of Class 66 locomotives, heads south through Doncaster.

By 2017, the company was stabling locos at Barnetby, on South Humberside. On 29 June 2017, their class locomotive, No. 70812, passes Doncaster on a light-engine move from Bescot to Barnetby.

Along with their Direct Rail Services' counterparts, Colas Rail's locos are often stabled in the sidings to the south of the station on the east side of the ECML. On 30 June 2017, no less than six locos were stabled. A pair of Colas Rail Class 60s, Nos 60076 and 60095, are seen alongside Nos 56096 and 56105. A pair of DRS's Class 66s, Nos 66422 and 66434, are to the right.

A similar joint occupancy of these sidings by a pair of DRS and Colas Rail locos occurred on 14 March 2019. Colas Rail's Class 56 locomotive, No. 56113, is seen alongside DRS's Class 88 locomotive, No. 88010 *Ariadne*. They are viewed from passing on an ECML passenger service.

Between 2014 and 2018, Colas Rail operated a fleet of ten Class 60 locomotives. Amongst their duties was the biomass flow from Tyne Dock to the power station at Drax. On 6 January 2016, a pair of these locos, Nos 60021 and 60076, head north from Doncaster to Tyne to take up these biomass duties.

Other Train Operators

West Coast Railway Company workings make occasional appearances at Doncaster. On 8 April 2012, for example, No. 57001 heads north through Doncaster's platform 4 on a northbound empty stock working.

Similarly, Royal Mail Class 325 units make occasional daytime appearances. On 24 July 2014, No. 325011 pauses at Doncaster as it heads north to Low Fell, Royal Mail's siding on the outskirts of Newcastle upon Tyne.

On 23 November 2018, the arrival of Rail Operations Group's No. 47815 caused excitement amongst the local rail enthusiasts. The Class 47 was operating a ScotRail High Speed Train power car move from Ely to Millerhill, near Edinburgh. Scotrail-liveried No. 43150 is seen immediately behind the loco.

Through the 2010s, RMS Locotec leased a dozen locos to DC Rail, or Devon & Cornwall Railways, a subsidiary of British American Railway Services. Their pairing of Nos 31190 (D5613) and 56311 are seen on a light engine move arriving at Doncaster on 9 September 2013. This Class 56 was later converted to Class 69 loco No. 69002.

Network Rail and Infrastructure

Network Rail operates a fleet of four former Class 37 locos, renumbered to 97301–97304, within the departmental number range. On 10 February 2017, two of them, Nos 97303 and 97301, are stabled in West Yard on a test train. The pair are sporting the familiar Network Rail yellow livery.

Network Rail's diesel unit No. 950001 is also a regular visitor to Doncaster. It is seen a month earlier, in January 2017, between duties. It was built in 1987, using the same bodyshell as the Class 150 'Sprinter' units.

Many of the Network Rail test trains operate with a Driving Brake Standard Open (DBSO) and one loco, rather than using a pair of locomotives. One of the fleet of fourteen such vehicles, DBSO 9701, is seen on 10 March 2020.

The yards at Doncaster are home to a number of snowploughs, stabled here for most of the year. ADB 965578 stands in West Yard on 9 June 2022.

The Doncaster area sees a wide variety of On-Track Plant visitors. On 27 January 2016, multi-purpose 'Railvac' machine 7095150030, purpose-built for vacuuming rail ballast amongst other uses, is seen between duties in West Yard.

Network Rail's DR98008 is used across the network as a video-recording unit. It is seen on 12 July 2017 in Marshgate sidings, adjacent to the Doncaster to Scunthorpe line.

Network Rail and Infrastructure

83

Plasser and Theurer's DR73946 was another visitor to the Doncaster area on 5 January 2018. Built in 2007, this machine has been operated by VolkerRail since autumn 2008.

Another VolkerRail movement through Doncaster is seen on 23 March 2023. Matisa tamper DR75402 is operating an out and back movement via Chesterfield and Toton.

A Variety of Visitors to West Yard and Wabtec

In previous chapters, we have looked at the passenger service patterns that occur at Doncaster on a daily basis. In addition to these, a wide variety of locos and rolling stock visit the rail workshops at 'Wabtec' and the adjacent West Yard. On 23 March 2023, for example, Chiltern Railways' Turbo unit No. 165004 stands inside the Wabtec complex.

In this aerial view of Wabtec on 9 June 2022, two veteran Class 47s are both looking in a very sorry state. On the left, Harry Needle's (HN Rail) No. 47703 still carries a badly worn Fragonset livery. To its right, sister HN Rail machine No. 47714 retains a faded Anglia turquoise and white livery.

A Variety of Visitors to West Yard and Wabtec

Over the years, almost all operators' units have been seen here. In this view on 22 August 2013, No. 153312 is seen in the Wabtec compound. The single-car unit was operated by Arriva Trains Wales at the time of this visit. It is about to leave and return to its home depot at Cardiff Canton.

Another single-car unit, No. 153382, stands in West Yard on 16 November 2001. The unit is about to return to its home in the South West of England, proudly promoting the 'Great Scenic Railways of Devon and Cornwall'.

Another Arriva Trains Wales unit is seen in West Yard on 16 April 2015. On this occasion it is Class 158 DMU No. 158834, which has been to the workshops for attention.

A wide variety of EMUs also visits Doncaster, including ScotRail's No. 320316, seen here on 29 June 2012. It has found a temporary home in bay platform number 2 before being moved over to West Yard and Wabtec.

London Overground's No. 321414, more usually found on Romford to Upminster duties during 2015, was another visitor to Wabtec in April 2016. The unit is seen just inside the works complex.

In a typical aerial view of the Wabtec yards on 3 September 2015, the variety of rolling stock within the workshops is evident. South West Trains' five-car EMU, No. 458508, is in the centre of this scene.

A glimpse of the yard at ground level is possible when arriving on trains from the north. On 25 October 2017, Southeastern Trains 'Networker' EMU No. 465919 is visible from a passing unit.

The workshops have also carried out work on ScotRail's pool of High Speed Train power cars. On 22 January 2019, Nos 43031 and 43152 are paired in back to back formation whilst standing in the Wabtec yards.

In British Rail days, almost 300 Mark IV coaches were built by Metro-Cammell to work alongside the Class 91 locomotives. Having passed into the privatised era, many of these coaches survive in operation with LNER today. On 14 September 2016, Mark IV coach numbered 12471 stands in Wabtec yard.

Proving that just about any rolling stock can be found at Doncaster, Tyne and Wear Metro unit 4018 was present on 23 March 2015. It is seen in West Yard being moved by diesel shunter No. 08401.

This view from the station's platform 8 on 2 August 2019 shows the variety of motive power that can be found in West Yard. LNER Azuma No. 800113 has just been taken out of traffic and is seen approaching. Stabled alongside in West Yard itself are DRS' No. 68008, DB Cargo's No. 67004 and LNER's No. 91108. Just visible behind is another DRS Class 68 locomotive No. 68009.

Our two final views of Doncaster's West Yard both feature two iconic locomotives that will be forever associated with the ECML. First, Class 55 'Deltic' diesel locomotive No. 55019 *Royal Highland Fusilier* is seen on 9 September 2013. Built in 1961/62, these twenty-two 'type 5' diesels were the mainstay of ECML workings for almost two decades, giving way eventually to the High Speed Trains.

Our second ECML icon is seen here in West Yard on the same day. Celebrity engine Gresley 'A4' steam locomotive, No. 4468 *Mallard* is receiving close attention. It holds the world speed record achieved by a steam locomotive when reaching 126 miles per hour on 3 July 1938. This remarkable speed was achieved about 75 miles south of Doncaster, between the villages of Little Bytham and Essendine, near Peterborough. Built in 1938, it was to travel the length of the ECML on express passenger services being renumbered 60022 in the British Railways era, before withdrawal in 1963.

This steam loco No. 60163 *Tornado* may not yet have achieved the same iconic status but it has certainly attracted a good deal of attention since its launch in 2008. It was the UK's first new-build steam engine for almost half a century. On 5 May 2017, it is seen stabled on Roberts Road depot.

Hitachi Movements on ECML

Hitachi Rail Europe has a railway rolling stock assembly plant at Newton Aycliffe in County Durham. Its Class 800 series units were put through their paces on this stretch of the ECML during their testing phase. Class 800 No. 800201 is seen heading north through Doncaster on 23 August 2018. It is returning to the County Durham factory following one of its test runs.

Similarly, the Class 800 units built for First Great Western were regular visitors, particularly in the early days of building and testing. On 20 September 2017, a pair of units, Nos 800004 and 800003, had earlier worked from Newton Aycliffe to Darlington. They are seen here on a working from Darlington to Doncaster IEP (InterCity Express Programme) depot and back.

The IEP depot at Doncaster Carr is situated to the east of the main running lines to the south. The site of the former steam shed and diesel depot, which had been there for 100 years, was cleared during the construction work. The depot carries out servicing and repairs to the Hitachi fleet. A glimpse of the depot is seen in this view on 30 May 2019, taken from a passing train.

To the North East and South West

A number of freight workings pass close to but not through the Doncaster station area. Some use a 3-mile freight-only line between Bentley Junction and Hexthorpe Junction. The regular workings include steel traffic between Hull and Masborough (Rotherham). On 25 October 2017, No. 60066 has just rejoined the Sheffield line at Hexthorpe Junction and is seen approaching Conisbrough.

Similarly, some workings approaching Doncaster from the west avoid the station by using the curve by St James' Bridge to reach the yards. On 1 March 2017, No. 66710 *Phil Parker BRIT* heads for Doncaster Decoy Yard on a rake of empty hoppers returning from Ratcliffe Power Station. The GBRf working is seen heading east through Mexborough station.

To the North East and South West

To the north-east of the city, the autumn railhead treatment trains (RHTT) operate over both passenger and freight lines in the area. In autumn 2017, a pair of Class 20 locomotives are on this Grimsby Town to Bridlington working, with No. 20305 at the front and No. 20312 on the rear. The ensemble is passing Hatfield & Stainforth heading west.

The longstanding bitumen flow between Lindsey, South Humberside, and Preston usually passes just to the north of Doncaster. This Colas working was in the hands of No. 60056 as it approached Hatfield & Stainforth on the return working to Lindsey on 28 October 2017.

LNER and InterCity Livery

What goes around comes around! A reduced number of Class 91 locomotives have been retained for LNER to use, primarily on services between London and Leeds and York. Twelve locos remain in traffic for the foreseeable future. The class carried an InterCity livery when built at the end of the 1980s. A variation of this colour scheme is now being applied to the remaining class members. On 10 February 2023, No. 91130 is seen calling at Doncaster whilst propelling a service to London King's Cross.